Snow

Miranda Ashwell and Andy Owen

Heinemann Library
Chicago, Illinois

Published by Heinemann Library,
an imprint of Reed Educational & Professional Publishing,
100 North LaSalle, Suite 1010
Chicago, IL 60602
Customer Service: 888-454-2279
Visit our website at www.heinemannlibrary.com

Printed and bound in Hong Kong
Designed by David Oakley
Illustrations by Jeff Edwards

03 02 01 00
10 9 8 7 6 5 4 3 2

Library of Congress Cataloging-in-Publication Data

Owen, Andy, 1961-
Snow / Andy Owen and Miranda Ashwell.
p. cm – (What is weather?)
Summary: Briefly discusses aspects of cold weather, including such topics as snow, frost, icebergs, blizzards, effects on plants and animals, and more.
ISBN 1-57572-790-0 (library binding)
Snow–Juvenile literature. [1. Snow. 2. Ice. 3. Cold.]
I. Ashwell, Miranda, 1957- . II. Title. III. Series: Owen, Andy, 1961- What is weather?
QC926.37.084 1999
551.57'84–dc21 98-42814
CIP
AC

Acknowledgments
The author and publishers are grateful to the following for permission to reproduce copyright material: B. & C. Alexander, pp. 5, 9, 16; A. Hawthorne, p. 22; Bruce Coleman Limited: p. 13; J. Johnson, p. 14; H. Reinhard, pp. 6, 20; K. Taylor, p. 4; M. Taylor, p. 22; S. Widstrand, pp. 18, 19; Robert Harding Picture Library, p. 23; N. Blythe, p. 28; L. Burridge, p. 7; Explorer, p. 29; J. Robinson p. 26; Oxford Scientific Films/C .Monteath, p. 8; B. Osborne, p. 15; Pictor International, p. 11; Planet Earth Pictures/S. Nicholls, p. 17; SIPA/Le Progres, p. 27; Still Pictures/T. Thomas, p. 12; Tony Stone Images/J. Stock, p. 24.

Cover photograph: B. & C. Alexander/The Stock Market.

Some words are shown in bold, **like this**. You can find out what they mean by looking in the glossary.

Contents

What Is Snow?

This snowflake is shown much larger than its real size. It was made in a cloud from tiny water drops. The drops were so cold that they turned to ice.

The ice became snowflakes. When snowflakes fall from the sky, we say that it is snowing. Each snowflake has a different shape and pattern.

What Is Frost?

The air is full of tiny water droplets. They can **freeze** on cold nights. The frozen droplets make a thin layer of ice called **frost**. Like snow, frost covers twigs and branches.

On very cold days, rivers and ponds can be covered with a layer of snow and ice. The ice will break where it is thin. You should never walk on ice.

Snow on Mountains

Over high mountains, the air is cold. Water drops in the clouds **freeze**. They fall as snow. The higher one goes up a mountain, the more snow he or she will see.

Over many years, layers of snow are crushed together. They turn into ice. The ice melts and freezes. It slides slowly down the mountain as a **glacier**.

The Coldest Places

North Pole

At the Poles, the sun's heat is spread out, so it feels colder.

Heat from the sun

sun

The sun never feels very warm near the North and South Poles. These places are the coldest places in the world.

Sunlight shining on snow is very bright. People on the snow have to wear sunglasses. They must also wear very warm clothes.

The North and South Poles

The South Pole is in **Antarctica**. Here, the land is covered in snow and ice. Penguins push together to keep warm.

The North Pole is in the **Arctic**. Here the sea has frozen into thick ice. There is no land. Snow covers the ice. Whales and seals swim beneath the snow and ice.

Icebergs

There are huge **glaciers** at the poles. They move toward the sea. Warmer weather makes the ice melt. Huge chunks of ice break off. They become **icebergs**.

Icebergs are mountains of ice. They float on the sea. Most of an iceberg is hidden under the water.

Frozen Ground

Land close to the poles is cold. The ground is covered in deep snow. People travel on sleds, skis, or snowmobiles.

Land near the South Pole is frozen hard. Only the top layer of the ground **thaws** during the short summer.

Animals in the Snow

Animals in snowy places have thick fur. Some animals have white fur. This makes it hard for other animals to see them when they hunt.

Some animals change color. This helps them hide from danger. These hares grow thick, white coats to help them hide in the snow. In summer, their coats are brown.

Plants in the Snow

Mountain plants have special ways of growing in cold, snowy places. These flowers have thick, furry petals to keep out the cold.

These trees have special, shiny leaves called needles. The needles do not fall off when it gets cold. They protect the trees from the cold.

Working in the Snow

Snow and ice can **freeze** your fingers and toes. People who work in cold, snowy places wear special clothes to protect them.

The Lapp people live among wild reindeer. They use thick, warm, reindeer skins to make tents, boots, and clothes.

Fun in the Snow

The best snowballs are made of wet snow. Wet snow is sticky. These snowballs hold together better.

Snow has a smooth surface, so it is easy to slide on. A sled moves very fast on the slippery snow.

Blizzard!

A big snowstorm is called a **blizzard**. A lot of falling snow causes problems. People in cars may be trapped when blizzards block roads.

Heavy snow and ice can break powerlines. Many homes will not have electricity. These homes will not have light or heat.

Snow Accidents

Snow at the top of a mountain can suddenly fall. This is called an avalanche. The falling snow picks up more snow. It sweeps down the side of the mountain.

Avalanches are very dangerous.
These men have rescued someone who was trapped in fallen snow. They had to work quickly to find this person.

It's Amazing!

More snow falls on Mt. Rainier, in the state of Washington, than any other place on Earth, more than 100 feet (30 meters) each year.

Snow fell in the Kalahari **Desert** in Africa on September 1, 1981.

Many **Inuit** people still build houses from blocks of snow.

The record for the most snow to fall during a 24-hour period is 75 inches (190.5 centimeters). It fell at Silver Lake, Colorado, on April 14, 1921.

Glossary

Antarctica land at the southernmost part of the world, around the South Pole

Arctic frozen area at the northernmost part of the world, around the North Pole

blizzard storm of heavy snow

desert place where there is very little rain

freeze when water turns to ice

glacier river of ice

iceberg mountain of ice floating in the sea

Inuit people who live in the coldest parts of North America, Russia, and Greenland

thaw to melt or become unfrozen

More Books to Read

Cech, John. *First Snow, Magic Snow*. New York: Simon & Schuster Children's, 1992.

Davies, Kay. *Snow & Ice*. Austin, Tex: Raintree Steck-Vaughn, 1995.

Riehecky, Janet. *Snow: When Will It Fall?* Chanhassen, Minn: Child's World, 1990.

Index